Copyright © 2005 North Amana Workshops

All rights reserved under International and
Pan-American Copyright Conventions.
Published in the United States by
North Amana Workshops.

ISBN 0-9761045-0-4

Manufactured in United States of America

About the Author

The author, Glenn H. Wendler, was born in High Amana, Iowa, in 1931, just prior to the reorganization of the old Amana Society. He received his education from the Amana schools and Iowa State University, from which he received a Bachelor of Science degree in Civil Engineering. A lifelong resident of the Amana Colonies, Mr. Wendler was an active volunteer and community leader. His many contributions to the Amanas included serving as an elder of the Amana Church Society for 40 years and serving as president of the Amana Society, Inc., from 1989 to 1995. He founded Wendler Engineering & Construction, Inc., in 1973. His wife and sweetheart, Mary, supported him throughout all of his life and in all of his endeavors.

As far as I can tell, the author, who was also my father, started this book shortly after his retirement from Wendler Engineering & Construction, Inc., in 1996. He collected photographs, researched history, and recorded his recollections over the next few years. He died suddenly and unexpectedly in 2001, prior to the publication of this book. Whether he actually finished the book remains a question. The book is organized by seasons and activities, and although the author's original writings span all four seasons, I do not know whether they include all of the various activities he intended to cover. One thing is certain: his original working manuscript designates the final chapter on the church as Chapter X, but there were only eight chapters found. Does this mean he intended to complete two additional chapters, perhaps on other enterprises or activities of the village of High Amana?

In editing this book, I have tried to keep the tone and spirit of the author's writing intact, choosing only to modify spellings and grammar in those cases where the unmodified version might otherwise interfere with the story.

Guy H. Wendler

Acknowledgements

The "Acknowledgements" section of a book is the place for the author to thank those who helped with the book. Due to his death prior to publication of this book, the author did not have a chance to complete this section. Therefore, to those who helped him with anecdotes, information, and photographs, we thank you on behalf of the author.

In addition, we would like to thank our aunt and the author's sister, Janet W. Zuber, for allowing her poem, "High Amana," to be included in the Conclusion; Marietta Moershel, Lanny Haldy, and Peter Hoehnle, for their review of the manuscript; Chuck Wendel, for providing the farm equipment photographs; and the Amana Heritage Museum, for providing various Amana photographs.

Guy H. Wendler
Stuart F. Wendler
Susanna Wendler

Contents

Preface	ix
Chapter I—The House	3
Chapter II—The Town	15
Chapter III—The Smithy	23
Chapter IV—The Farm	31
Chapter V—The Sawmill	59
Chapter VI—The Carpenter and Wagon Shops	67
Chapter VII—The Store	75
Chapter VIII—The Church	89
Conclusion	107

Preface

Now that I am "middle-aged" (actually, 65 years old), I have determined that all things remain relatively the same for about three generations. One remembers things as told you by grandparents. You cherish the ways your parents raised you, the love they showed, and the values they instilled. And finally, you believe that what you stood for in the more mature part of your own lifetime has some value and is, therefore, worthy of perpetuating.

In reality, however, things are always changing. Nothing remains the same, even though memories tend to make you believe otherwise. Having come to that realization, I thought it proper to pass along some of the cherished memories of my childhood. Particularly so, because I was born and raised in a singularly special kind of environment, one that I doubt will ever be found again anywhere else in the world.

I was born on October 26, 1931, in a little village in east-central Iowa called High Amana. This village was first known as *Amana vor der Höhe* (Amana in front of the heights), and is one of the seven villages that constituted the Amana Church Society. The Society was the home of a group of about 1,100 German and Swiss immigrants who came to America in 1842 and first settled in a community near Buffalo, New York. Ten years later, they began relocating in the Iowa River Valley in Iowa County, Iowa, 20 miles west of Iowa City.

The founding philosophy of the Society was based upon certain Pietistic Christian religious beliefs. The main distinction from other Protestant churches was the belief of its members in the Word of Inspiration which is received from God through *Werkzeuge* (individuals who were His "instruments"). That belief is still a distinguishing tenet of the church as it exists today, although there have not been any *Werkzeuge* since the late 1800s.

The people of the pre-1932 Amana Society lived as a religious commune. That is, they practiced communism in the real sense with each person working for the common good according to his abilities and each receiving food, clothing, shelter and other necessities of life in equal measure.

The communal aspects of the Amana Society were abolished by a vote of the members in 1931, and final separation of the church from the economic matters of the organization came about in 1932. Like all communistic organizations, sooner or later the members determine that not all are being treated equally or that they need not contribute to their full ability since they will still receive an equal share. The economic pressures which contributed to the breakup of eastern European communism in the last decade of the 20th century are not unlike what happened in the Amana Society. It is interesting to note that the 19th century communism of the Amana Society and that of the 20th century eastern European nations had almost identical life spans.

Much has been written about the Amana Church Society. The reader is advised to consult the archives of the Amana Church in Middle Amana, Iowa, or the Amana Heritage Museum in Amana, Iowa, for further information.

Prior to the breakup of the old Amana Society in 1932, there was one last child born under the old system. That was me.

The Commune's Last Child

Chapter I—The House

I really disliked going to bed at night. For me it was just another eight to 10 hours of wasted time lying there doing nothing except sleep, until finally I awoke to the sounds of the new day coming through the open windows of the house.

The house was a big sandstone structure located on the north side of the street, across from the general store. It was built in 1859 with massive hand-hewn blocks of sandstone hauled from a quarry located at the bend in the road to Middle Amana and from another quarry in West Amana. The house was not unlike a musical instrument.

One had to learn to play it to get the most out of it, not only from a child's point of view but also from a practical one. Its residents made subtle changes to its physical characteristics to compensate for the wide swings in seasonal climatic changes known all too well by Iowans.

The house had no insulation as measured by today's standards. The plaster was bonded directly to the inside face of the sandstone blocks. Tiny voids of air between the grains of sand helped keep out winter's bitter cold. The wind whistled underneath the windowsills and through the "nine-over-six" (that is to say, nine panes in the upper section and six in the lower) double-hung windows, a peculiarity of Amana architecture. The house had two stories—four if one counts the basement and attic.

The house, circa 1930; Speisesaal *at left and town water tank (with conical top) at upper right*

My family lived upstairs until *Geiger Opa* passed away. The Geigers were not really my grandparents, but we endearingly called them *Opa* (grandfather) and *Oma* (grandmother) because they were older and always around when help was needed in the care of this little boy. Even though I did not get to know them well, everything I remember of them is good. *Geiger Opa* was an old man when I was born. He was an elder of the church and, from what I recall, a kind and gentle man. I remember his funeral. It was the first one I ever attended, and it left quite an impression upon a five-year-old. After he died, *Geiger Oma* moved to Homestead to live with her daughter and family. It was at that time that we moved downstairs.

The sandstone part of the house was very simple in construction. It had four big rooms on each floor with an entrance hallway in the center. The stairway to the second floor was unique insofar as Amana architecture was concerned. It made a quarter turn on its way to the second floor, a characteristic that was maintained when my parents remodeled the interior of the house in the late 1940s.

My parents and I shared a large bedroom in the northeast corner of the main house. We had a "Sunday room" which was used only on special occasions, such as Christmastime, funerals, and when church elders from the other Amanas visited my folks on special church service days. This room contained all the "fancy" furniture, most of which was built by craftsmen in the Amanas. Some items came from Germany with Amana's first settlers and had been lovingly cared for and handed down through several generations.

In one corner of the Sunday room stood the fancy wood-burning stove. It was "fancy" in that it received better care and had an ornate design on its castings. Some Amana stoves even had a chromium-plated footrest at the bottom to warm toes on cold evenings, but ours did not. The stove stood about four feet tall. Its base was made of cast iron shaped in an oval. The firebox was the main section of the stove. It was made of thin steel formed into an upright, oval-shaped cylinder. The top was a casting and riveted to the firebox. The door for the wood was on one end and on the front side was a device called *der Schieber* (the slider), which regulated the draft and controlled how hot the fire in the stove would burn.

The top of the stove had a large hinged casting so that a larger piece of firewood could be laid on the fire. A blue-black, vertical stovepipe rose upward toward the ceiling and then across the room to one of two chimneys, venting the smoke into the outside atmosphere. On a snowy, moon-lit night nothing was more picturesque than an old Amana Colony street scene with smoke gently winding upward from the chimneys of the houses.

Earlier picture of the house, probably taken in the late 19th century

The windows of our Sunday room were dressed with lace curtains and green Rollos, or pull shades. These served as an integral part of the cooling system, which I will explain later. A large table stood between the windows on the south wall and a library table on the opposite wall. I never understood why it was called a library table since we did not even know what a library was. An Amana couch (the "fancy" one) stood in another corner mingled with mismatched straight-backed chairs and a few nicer chairs covered with needlework seat cushions. The carpeting was crafted locally on a manual loom and designed by the women of the house using the colors from strips of rags carefully saved from worn-out garments and other pieces of cloth.

The carpets were of simple design made by alternating strands of different colors or materials. That is not to say that no planning went into the selection of the pattern. To the contrary, the colors and materials available to the housewife had to be weighed in order to provide a recurring pattern which fit the size of the room and matched the Amana Blue walls (two colors prevailed—Amana Blue and plain white). Here and there, in the more traveled portions of the floor, one found scattered braided rag rugs made by our grandmothers and mothers. Our Sunday room was a cozy room, particularly when filled with visiting members of the family on special church holidays.

Today we are accustomed to having electro-mechanical air conditioning systems in our homes. The householders of old Amana had their own type of "air conditioning" system which sufficed for many years. Iowa residents are well aware of the extreme climatic conditions during the year. High humidity and temperature differentials from minus 30 to plus 105 degrees Fahrenheit are not uncommon.

This is how the natural "air conditioning" system worked. First, most houses had wooden trellises built on the exterior walls of the structure (see picture on page 7). Grapes grew on these trellises, and the large, thick leaves provided shade so that the hot summer sun did not shine directly on the stone or wood walls of the houses. The massive sandstone blocks did not give in easily to the changes in temperature, but once doing so, they just as reluctantly gave up either the hot or the cold. Even the wood-sided frame houses had liners of light, straw-filled brick between the studs to provide some insulation. The grape trellises definitely kept the house cooler when the summer sun beat down.

Beyond the natural cooling provided by the shade of the grape trellises, another method for cooling was utilized. It was up to the housewife to regulate the Rollos and windows to take advantage of every possible manner in which the house could be kept cooler. During the nights, the windows were opened to let the cool air flow into and through the house. During the day, as soon as the sun got higher, the windows were closed and the shades were drawn to keep out direct rays of the sun. If a summer storm happened to approach, the shades and windows were opened to let the cooler air in that generally accompanied these storms. In the wintertime, every air leak in a window or doorjamb was plugged or covered. Most windowsills were provided with stuffed, sausage-like pillows, about two inches in circumference and as wide as the window, to keep drafts from coming through the crack between the window and the sill.

The other large room was our everyday living room. It held my dad's massive leather rocking chair, a large round oak table that served as a dining table on family gatherings, some "odds-and-ends" chairs, and a couch. Electricity came in 1936, but until that time coal oil (kerosene) lamps provided the light. The floor was covered with linoleum. In one corner stood another wood-burning stove with firewood for the day piled into a woodbin along the wall behind it. After electrification, our prized Howard brand radio sat on a south windowsill. It provided hours of entertainment for the family.

Next to the everyday room, in the northwest corner of the sandstone structure, was the kitchen. This was the only room to have running water (cold only). The white cast-iron sink with drain board served for food preparation as well as general hygiene—probably not the best combination, but it was all we had. Water was heated on the wood-burning stove during the winter and on the kerosene cooking stove in summer. Water for bathing was an entirely different matter with a later story.

Our food was cooked on a four-burner kerosene stove. There was always a distinct odor of kerosene. (Looking back, I have trouble understanding why this constant odor did not make us nauseous!) For baking, mom had a portable oven that was placed directly on two of the stove's burners. (Some houses had a round oven built into the stovepipe above the wood-burning stove, an ingenious device capturing the heat which would otherwise escape up the chimney.) One small freestanding kitchen cabinet held the provisions and the "everyday china." The "good china" was acquired a piece at a time from the special sales promotion coupons found in the product boxes of Quaker Oats Company products. Mom's "silverware" was similarly acquired through the Betty Crocker promotions of the General Mills Company. I remember how very, very proud she was when a new piece came through the mail and how lovingly she added it to her collection.

I was born (I am told) in the upstairs, northwest room of the house. My birth was attended by a midwife, *Schwester* (sister) Lina Keller, and by Dr. Christian Hermann, who had traveled by Model A from Middle Amana two miles away. Surprisingly, I do not remember my sister's birth at all. She was born five years later, downstairs I presume. I was probably shuttled to my Wendler grandparents or to an aunt and uncle in Middle Amana. I had no notion of what was going on except to find this new baby girl living with us who seemed to get all the attention from everyone.

This pretty well describes the living quarters of the house. However, the whole complex was much larger than this. Originally, it had been built as a communal kitchen house. As such, it served as a central kitchen and dining facility for a number of families in the pre-1931 High Amana community, as well as for various day laborers who worked the farmland surrounding the village.

Attached to the sandstone part of the house were several wood-frame structures. One contained the large kitchen where the communal meals were prepared. The kitchen was complete with a wood-fired brick hearth. This hearth was waist high and had a massive cast-iron cooking top and built-in convection oven. In yet another part of the frame addition there was a *Speisesaal*, or great dining room, with several large serving tables and benches for seating. Also attached was a woodshed and *Wäschhaus*, or washhouse, with an attached stone hearth baking oven. In one portion of the woodshed/washhouse complex was a rather elaborate (for those days) privy. Other unattached buildings included a chicken house, garden house, and brooder house. The whole structure had a tremendously large basement and attic with secret places that could keep a young lad occupied for rainy days on end.

The *Wäschhaus* constituted an important part of our residence. In the one corner there was a masonry hearth that held a large cast-iron cauldron. Early each Monday morning, Mother (sometimes Grandfather) pumped water from the cistern into the cauldron and built a fire in the hearth underneath to heat the water for the laundry. The laundry was done in a hand-operated wooden washing machine, run through a hand-operated wringer, and then hung outside to dry. On cold winter days, the union suits (underwear) would freeze and look like ghosts as they flailed in the wind.

The same *Wäschhaus* preparations took place on Saturdays, except the water was used for a different purpose. It was time for our weekly baths. (Remember, those were the days before hot running water.) Mother would ladle the hot water into a wooden tub and mix it with the cold until it was just right for my sister and me. I still remember the vigorous scrubbing we got. The *Wäschhaus* was a steamy, cozy place during the winter months, but oh my, how cold it was to run back to the house when the bathing was done!

The main house was connected to the woodshed and washhouse by a large, overhanging L-shaped porch that on two sides surrounded a courtyard-like area. A hand-dug well with a green cast-iron pump was centered in the concrete-paved courtyard. Ladders for maintaining the buildings hung under the porch roof on the outside wall of the woodshed.

The land surrounding all these buildings was used primarily for vegetable gardens and, in the fall, bundles of drying herbs and onions hung on wrought-iron nails driven into the porch rafters. From the corner of the courtyard, the pungent odor of new potatoes, the acrid aroma from crocks of fermenting sauerkraut, and the subtle bouquet of new grape wine wafted through the open cellar door.

The house was a wonderfully exciting place in which to grow up. It fired the imagination of a little boy and provided untold hours of play and innovation. It was a place where conjures of the mind were more important and longer lasting than all the modern toys and paraphernalia of today.

Chapter II—The Town

Amana vor der Höhe, or High Amana, is one of seven small villages comprising the Amana Colonies in east-central Iowa. Like its six sister villages (Amana, East, West, Middle, South, and Homestead), this village was basically a self-sustaining community. Even though the population was less than 100 people, the town had all the necessities to sustain itself. This made it a place of untold excitement and enterprise for a young lad.

On a given day, one could see logs being sawed into boards at the steam-driven sawmill, wagons

Amana vor der Höhe *(High Amana) circa 1900, but much as it still looked in the 1930s*

built in the wheelwright shop, horses shod at the smithy, bread baked at the bakery, shoes repaired at the cobbler's shop, and hogs slaughtered at the butcher shop. Added to all of this were the ever exciting times when the delivery trucks from Cedar Rapids made their weekly stops at *Onkel Willie's* (Uncle Willy's) general store across the street from our house.

One of my grandmothers, *Wendler Oma* (Susanna Haldy Wendler), told me of her mother's first memory of coming to High Amana. She came to High Amana via the first settlement on the Canadian border near Buffalo, New York—a place named Ebenezer. At the time she arrived in High Amana, only the church and a few houses had been built. A creek ran north and south through the center of the town just east of the church building. The creek, benign for the most part, could develop into quite a torrent during severe storms. Two bridges for horse travel crossed the creek along with a few foot bridges allowing access to either side of the little creek. The surrounding area was covered with prairie grasses and hazelnut clumps. It was a virtual wilderness.

As noted, this small creek created quite a problem for the residents in the early days. Flash floods exceeded its banks and often wiped out the gardens nearby. The elders solved the problem by re-routing the ditch to the west of the little town. All the work was done with oxen hitched to hand-operated scrapers and ox-drawn, two-wheeled *Schnapp Karren* (dump carts) holding a cubic yard or so of material. The quarter-mile project diverted one drainage area into another one and, estimated conservatively, required the excavation of around 10,000 cubic yards of earth. To this day, this "new" ditch is still known as the *Ochse' Grabe* (Ox Ditch).

A total of four large communal kitchens were built in High Amana. Ours was known as *Geiger Küche* (Geiger's kitchen). Directly east of the church (now the Amana Arts Guild) and across the ditch was *Hetz Küche*. South, at the corner of the present Iowa Highway 220, was *Pitz Küche*, and the middle house on the west street of the town was *Haldy Küche*. My paternal grandmother was born in the residence connected to Haldy's kitchen and lived there until her marriage to *Wendler Opa*. Each kitchen was responsible for serving about 25 people. The use of the kitchens, as such, was discontinued in 1932 when the communal system of the Amana Society was abolished. The kitchen and *Speisesaal* in our home remained unaltered for a number of years into my childhood. But in 1935, as the rural countryside was electrified by virtue of the Rural Electrification Act, it was transformed into a "show room" for a modest variety of light fixtures soon to be installed in the homes of Amana. My father, Harry Wendler, was a member of the "wiring team" assigned to this big undertaking.

Interior of typical pre-1932 communal kitchen

One of my boyhood friends lived in *Hetz Küche*. His family was large and shared the house with another family of relatives. Because of the many mouths to feed, the old kitchen and the *Speisesaal* were utilized during the summers of the early 1930s. Occasionally I was invited to eat *Mittagessen* (lunch) there, and so I did get a flavor of how the old communal kitchen operated. The diners sat on plain wooden benches around a long dining table. The food, prepared on the old hearth, was brought from the kitchen in large, heaping bowls. It was a neat, plain, and friendly place with conversations going on about the field work, carpentry, and gardening.

The map on page 21 provides a layout of the town as it was in the early 1930s and gives the reader a mind-picture as to the location of the various enterprises which we will visit as we relive some of this little boy's exciting and fun-filled days back then.

Perhaps the best way to enjoy the sites and enterprises to be found in my childhood town of High Amana would be to do it in the order of the four seasons, beginning with spring. Unlike today, seasonal influences had a tremendous impact on the lives of the early settlers of this village. The farmyard in the spring and summer presented a picture totally different than in the fall or winter. Agricultural chores were entirely dependent upon the time of the year, especially in High Amana, where there were no large enterprises such as the woolen mills of Main Amana and Middle Amana, the furniture factory in Main Amana, or the flour mill in West Amana. Nonetheless, High Amana did have a myriad of interesting enterprises necessary to keep the farm and the kitchen houses operating.

The town, later in 1948

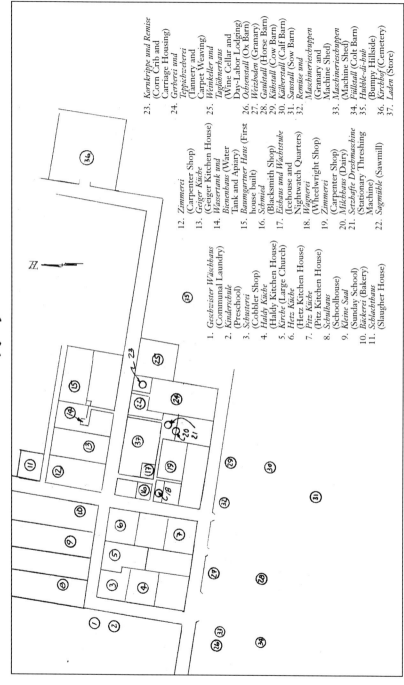

Chapter III—The Smithy

The end of winter meant a lot of work for my *Wendler Opa*. He was the village smith, the man in charge of shoeing the dozen or more teams of horses needed to keep the farm operating. During the winter he prepared the summer horseshoes to replace those which the horses wore during the icy winter months.

Let's take a look inside the smithy. It had a large double door facing the alley to the south (the lot, 708 13th Avenue, now has a tan brick house situated on it). As we walk through the door we would see the forge immediately to our right.

Around the forge and on the walls behind hung all sorts of tongs with long handles and of special designs to grab the various thicknesses of steel needed to make and repair the many special tools and machines used in the village enterprises.

The picture above shows the author's Opa *and* Oma, *Arno and Susanna Wendler, posing in the front yard of their home across from the smithy. The building behind Arno is the shop where he served as the village blacksmith.*

Further to the right, behind the chimney, was the large bellows which provided the air for the forge. I recall it was about six feet in diameter and, though much larger, shaped almost like a giant fireplace bellows. The bellows was actuated by an overhead system of levers, pins, and yokes connected to the back and top. A long wooden lever extended outward to where the blacksmith worked the forge. The end of this lever arm was positioned so that the blacksmith could reach up and pump just the right amount of air into the coals of the forge.

At the end of the wooden arm, polished brightly by the blacksmith's rough hand, was a brass knob like one finds as a decoration on a harness hame. My *Opa*, puffing on a pipe full of tobacco, would gently pump the bellows to provide just the right amount of air and oxygen needed to heat the steel to the correct temperature. With a long pair of tongs he would poke the fire and turn the steel until it glowed fiery red. He would remove the hot metal from the fire with his tongs and then place it on a large anvil where he shaped it by pounding with a heavy hammer, re-heating the metal, and then shaping it some more until it became the object he had earlier envisioned.

The smithy was a warm place to be in the late winter and the early frosty days of spring. The sparks would fly as the white-hot steel was struck. The forge fire always kept the place cozy. By spring enough new horseshoes would be hanging to shoe all the horses in High Amana.

The teamsters brought their teams to *Opa* in the morning and generally stayed to help with the shoeing. *Opa* would select a set of shoes that best fit the horse—some were larger, some smaller. The old shoes were removed from the hooves with a large pair of pinching tongs. He would clean the horse's hooves and shape them by giving them a manicure with the help of a very large, rough file and a hooked hoof knife. The old horses stood calmly while the shoeing took place, but the young ones were a different story. They were tense and restless, and became quite agitated by the process. Even though the spring days still had some of winter's frostiness in them, the blacksmith and his helpers often ended up in sweat-stained shirtsleeves before the young horses were finally shod. Coincidentally, that was often a young boy's first lesson in hearing words which were not to be repeated in more genteel surroundings.

Once the hooves were trimmed and shaped, the smith took the selected shoe and heated it on the forge. Then, with another special shaped pair of tongs, he placed the hot shoe against the horse's hoof. It burned and melted the gelatinous hoof so that the shoe fit perfectly. The smell of burning horse hooves was another special odor of the village—pungent, but not necessarily unpleasant. A good horse man or woman can recall the smell at will. The shoe was then doused in cold water and finally attached with new shoe nails, eight to a hoof. The nails were inserted through pre-made holes in the shoe and then driven through the outer portion of the hoof, the part which has no feeling. The nails where then clinched and clipped with the final touches made with a large rasp or file. And so it went for a week or more until all the teams had new summer shoes.

There were of course many other duties for the blacksmith during the year. Sled runners were fashioned for the large horse-drawn sleds used on the farm during the snowy months of the year. Rims for the large wooden wagon wheels had to be fabricated along with various metal hardware items necessary to keep the farm and other shops of the village operating.

Some of the work required the use of a large drill press driven by a one-cylinder, five-horsepower gasoline engine. The engine drove a belt connected to an overhead system of line shafts, pulleys, and belts which in turn powered the several machines in the shop. I remember a belt-driven grinder, a drop-forge hammer, and a metal saw in addition to the drill press. I also remember a corn sheller and feed grinder for chickens raised in the hen houses behind each kitchen.

Outside the smithy was a thick, flat, circular piece of concrete about five feet in diameter with a hole at the center. Also in the front yard was a concrete trough, about six feet long and a foot wide. At the center of the trough and on each side was a steel pipe imbedded in the concrete. Each pipe was about three feet high and had spikes welded at equal spaces, parallel to the trough. I will try to explain for what and how these two structures were used.

Every so often the rims of the large wagon wheels needed maintenance or replacing. Sometimes a complete new wheel was required. In any event, it was the smith who made and installed the new steel rim. He would measure the outside circumference of the wheel with a rolling wheel-type of measuring device and determine the exact circumference of the wheel in question. Then he would fashion a steel rim out of a strap of metal, roll it into a circle, and finally weld the ends together on the hearth and anvil. The new rim would be just a bit smaller in diameter than the wheel itself.

Once the forge-welding of the rim was completed, it was again heated in an outside fire pit so that the metal expanded just enough to allow it to slip over the spoke wheel now laying on the circular concrete slab. Once in place, water was doused on the rim and it shrank tightly onto the wheel. The wheel was then moved to the spoke supports on the trough described earlier and now filled with linseed oil.

The wheel was slowly spun to give it a protective soaking of oil. Even the older wheels were regularly taken from the wagons to receive the oil bath.

There was at one time a tin smithy in the village of High. This shop was responsible for the manufacture of tubs, kettles, scoops, and other utensils used in operations of the kitchens and the farm. High Amana no longer had a tin smith by the 1930s, but some of the other Amana villages maintained such shops into the 1940s.

Chapter IV—The Farm

The first thing that comes to mind about the spring farmyard is the mud. At night the frost would make the ground hard, but as the day progressed and as the horses and men moved about doing their chores, it became a knee-deep quagmire.

As the last snow melted from the unsheltered places in the farmyard, the farm became a beehive of activity. Wagons had to be converted from bobsleds to the wheel-type running-gears. The farm boss had to estimate the number and types of wagons that would be needed first—hay wagons, box wagons, etc. Then, with the help of the wagon maker and wheelwright, all were checked for bad wheels, cracked reaches, missing hardware, and broken double trees.

Wheels were removed and inspected, and the hubs greased. A simple wheel jack was used to lift the axles of the wagons off the ground so the wheels could be removed. The wheel jack was a trapezoid affair made of five pieces of wood. With the eccentric lever in the upward position, the jagged-sloping top piece slid under the axle of the wagon. The lever was then pushed downward utilizing an eccentrically-actuated force to lift the axle. The large nut holding the wheel to the axle was removed, and the wheel came off. Thick, special grease that came in 25-pound buckets was daubed on the axle. The wheel was replaced and ready for another summer's work. Other farmhands were getting plows, oat seeders, and corn planters ready.

Yes, spring was a time of active frenzy on the farm. Horse harnesses, complete with the fly nets, were repaired and oiled for the summer. I can still see the dipping tank used for oiling. It looked like a giant, green upside-down top hat. You almost expected to see a giant magician pulling a rabbit from it. It was filled with harness oil to about half its capacity. Once the harness had been repaired, it was thrown into the tank, stirred, pulled out, and laid on the rim of the "top hat" to drain the excess oil back into the tank. I believe the oil was heated somewhat for better penetration into the leather.

Spring brought with it another event which was always a risqué attraction to the youngsters in town. It was the breeding of the mares. Looking toward West Amana we would see Jakob Jaeger's buggy coming down the road pulled by a nondescript, but pretty, gelding. Tied to the tailgate of the buggy was a magnificent, white-blazed, chestnut stallion. He must have stood 18 hands high or more. Mr. Jaeger had him groomed to perfection, sometimes with ribbons braided into his black mane and tail. The stallion's four, white-stocking legs reflected the morning sunlight as the trio approached the town. *Boy-oh-boy, what a day this was going to be! If we can just get home from school in time!*

Jakob Jaeger, by the way, was a freelance operator who lived in a small, rented apartment in the hotel in Upper South Amana. He was not a member of the Society, but made his living by furnishing stud services for all the Colonies' mares, as well as those of private farmers in the area.

Spring meant the start of new life all around us. The colts and the calves were being born. My *Wendler Opa*, in addition to his duties as smith, was also *der Füller Bass*, which meant that he was in charge of overseeing the births of the colts and tending to their care (*der Füller Bass*, pronounced something like "the filler baas," means colt boss, or keeper of the colts). I never really got to see a colt born during the time *Wendler Opa* took care of them, but it was great fun to see them trying out their new legs and searching for the first taste of mother's milk.

Now the snow was finally gone from even the most protected places, and the teamsters with their implements went into the fields. The plowmen came first with their teams of four horses. I barely remember this because shortly after the 1932 change from communal life, gasoline tractors were introduced on the farms. Nevertheless, most of the farm chores in the early 1930s were still done with horses.

After the plowmen turned the fields, other teams with discs and harrows prepared the soil for planting. The man who was in charge of actually planting the corn was considered an especially talented person. A lot of pride went with the teamster able to create "straight-as-a-string" rows of corn.

Corn was planted with a planter that was actuated with a checking chain or wire. The check wire was as long as the field, and some fields were quite long. The wire itself was about No. 9 gauge and had knots of wire wrapped at equal

Planter

intervals along the main wire. The check wire was staked down at each end of the field, and before the planter proceeded across the field the wire was first placed into the checking wheel of the planter. The little umbrella-like disk above the seat of the planter, as shown in the above picture, was lowered. This disk marked a groove into the ground, giving the teamster a center line to follow for his next turn across the field. As the planter started across the field, the knots on the wire tripped the planter mechanism and, depending on what template was in the planter, the appropriate number of kernels of corn were dropped into the furrow. At the end of the row the operator had to turn around and proceed in the opposite direction. The trick for him was to be able to move the wire and its anchoring stake so that when the plants emerged from the soil in a few weeks it looked as if they were equally spaced in rows in both directions, across the field as well as in the direction the planter traveled. This allowed the farmer to cultivate the new corn for weeds, not only the direction in which the planter planted, but also crossways. You could always tell a good operator if his rows "cross-checked," and it was a proud teamster who looked down new corn rows seeing that his work was perfect in all directions.

Once the new corn plants emerged, it was a battle to keep the weeds down. There were, of course, no herbicides, and since the only fertilizer was the good old stuff from the manure piles, weeds grew abundantly. The horse-drawn cultivators were generally one- or two-row affairs. Each had shields which ran close along the new corn shoots to keep them from being covered with the soil plowed up by the cultivator shovels just outside the shields. Later, when the corn was a foot tall or so, the shields were removed and the corn was "laid-by," or "hilled." We kids would often search the little tool boxes on the cultivators to look for Indian artifacts which the teamsters might have picked up during the day as they endlessly looked at the corn passing beneath the seat of the machine.

Another important farm operation took place at the same time the corn was being cultivated.

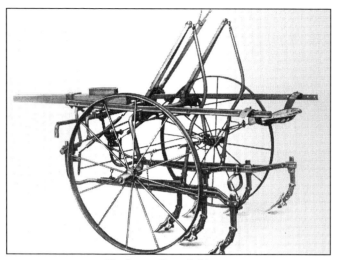
Single-row Corn Cultivator

The first crop of hay was now ready for mowing. This was done with a horse-drawn sickle mower. The two wheels of the mowing machine were made of steel. One wheel had a gear device mounted on the rim or spokes. This gear drove a shaft to a gear box, which powered an eccentric bar connected to the sickle bar itself. This was the pitman arm. It moved the sickle back and forth between stationary teeth.

The sickle bar slid along the ground and cut the hay as the sickle went back and forth. At the very end of the eight-foot sickle bar was a wooden blade which separated and laid back the newly-mown hay from that still standing. The result was a clean, narrow path between the mown and unmown hay. It is virtually impossible to describe the sweet fragrance of newly-mown hay to someone who has never had the opportunity to smell it.

Mower

The hay was allowed to dry for a day or two, after which someone with a dump rake would come and move the mown hay into windrows. I suppose the best way to describe the dump rake would be to imagine a giant garden rake. The handle of the rake would be the tongue for the horses, and the metal part of the rake itself would have a wheel on each end. As the horses pulled the rake across the hay, the hay accumulated. Every so often, the operator would lift the rake to dump it and then resume raking until it was full again. The dump rake had a seat for the teamster who controlled the horses so that they walked a nice straight line across the mown hay. When the rake was full of hay, the teamster stepped on a pedal, actuating a wheel-powered mechanism that lifted the tongs on the rake which left a pile. The trick for the teamster was to dump the rake at precisely the end of the windrow of his previous round. If done correctly, the end result would be neat, straight, and continuous windrows.

Rake

I became a teamster at the very young age of 11. The farm manager (my *Onkel Heinrich Bendorf*) told me to hitch up the dump rake and go rake a field called *das hundert Acker Stück* (the hundred-acre field) to the west of the village. Well, I hitched up and, with *Onkel Heinrich* following in an old Ford pickup truck, I headed for the field.

Once there he gave me instructions on how to operate the rake and described how the field should look once I was done—neat, straight, and continuous windrows of hay. Well, for one thing, the foot pedal that actuated the dump mechanism was not the easiest to push for an 11-year-old, notwithstanding the fact that it was virtually impossible for me to reach while sitting on the seat.

My team of horses was old and slow. One was named Queen and had been a horse in a team driven by my father years earlier when he had worked on the farm. The field was large and the job became boring quite early in the day. Sometimes I would forget to dump the rake at the appropriate time, and even when I remembered, I usually had a heck of a time actuating the dump lever. I can still see the face of *Onkel Heinrich* when he came out to inspect my handiwork, wondering how those randomly scattered and variously sized clumps would ever by loaded into the wagons. Luckily, the dump rake was soon replaced with the side delivery rake, which required less brains and less brawn.

After the hay had dried in its windrows, the time came to move it into the barns. By this time some of the wagons had been rigged with hayracks. The hayrack was a larger-sized wagon platform. It measured about eight feet wide and

15 feet long. Most of them had semi-open sides with front and back side boards to contain the hay. A ladder in front allowed the teamster to climb up and down the load.

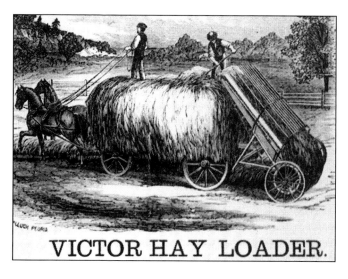

VICTOR HAY LOADER.

A mechanical pull-type machine, as shown in the picture above, was utilized to load the hayrack. The team, wagon, and loader straddled the windrow of hay. As the loader passed over the windrow it picked up the hay and with a series of pronged arms moved the hay up a metal chute. The arms were so arranged that when one moved inward and up, the next moved outward and down, thus grabbing and letting go of the hay as it moved up the chute of the loader. At the top of the chute, the hay dumped onto the rack and was stacked by the teamster. Once fully loaded, the wagon was unhooked from the loader and taken to one of the many barns in the farm complex.

What hay went where was determined by what animals were kept in the particular barn. Alfalfa went to the dairy barn, and timothy hay went to the horse barn. Hay of a lesser quality went to the steer and sheep barns. I vaguely remember High Amana having a sheep herd—the barn was located on the hill just north and east of where the butcher shop was located (see map on page 21). The following picture gives an idea of how the hay was hoisted from the hayrack with a rope-and-pulley system and then dropped into the haymow of the barn. Usually, three or four men were required to level the hay once it dropped from the fork.

My very first job on the farm was leading the pull horse. (Note: In High Amana, we did not ride the pull horse, as pictured on the opposite page, but instead led it.) The teamster set the hayfork. He would then give the signal, and the pull horse with the lift rope attached was led away from the barn. This moved the loaded fork upward toward the peak of the roof where the catch trolley was located on a rail suspended under the peak. Once in the "catch," the load moved horizontally to the place where the stacking men wanted it dumped. They yelled, the horse was stopped, and the trip-rope man dropped the load of hay. As the horse returned to its starting place, the trip-rope man pulled the fork back for another load while another boy pulled the big rope back towards the barn. Two of us young boys would alternate pulling in the rope and leading the horse. Workdays were often 12 hours, and it was pretty hard work for a small eight-year-old.

The other thing that made it tough for a young lad was the fact (and I do think it was done with forethought) that the largest, oldest, and clumsiest horse was designated as the rope horse. I remember my first one. His name was King, as I recall. He stood about 100 feet tall, had hooves that measured 12 inches in diameter, and weighed at least a ton—anyway, that is how he appeared to me. The other distinguishing fact about the horse was that he had the uncanny knack of placing one or the other of his humongous feet precisely upon those of an eight-year-old boy and then completely forgetting to make another move, even after hearing the epithets of a young boy in misery. The words first heard at the smithy came in handy at that time.

Can you imagine the energy it took for a boy of my age to harness this brute every morning? The collar itself weighed 30 or 40 pounds, and it took two of us—one on the ground and the other standing in the feed box of the horse's stall—to slip the collar around King's head every morning. Then came the tangled mess of a harness. Often we did not even know which end went where. When we finally had the old "plug" harnessed, the two of us stood there, dripping in sweat, and the first hour of a long day was almost over. Every once in a while a kindhearted old teamster would come over and give us a hand, but only after giving us ample time to become completely exhausted.

At least in the springtime it rained often enough to give us breaks in our steady pace of work. In midsummer, during the second and third crops of hay, it became almost unbearably hot, and the days dragged on, one after another. The poor old guys who worked in the haymow under the tin roofs of the barns really suffered. I am sure that it was only the bottle of beer every two hours or so that made it possible for them to continue. Even so, everyone seemed rather happy and content. At the end of some especially long days the whole farm crew would jump into a wagon and head for the Mill Race for a swim.

Every once in a while the thick, long hay rope broke. This gave us a respite from our labors while we waited for old Mr. Murbach (he worked in the carpenter shop) to splice the rope. Only a few men knew how to do that, and he was one of them. The whole process could take quite a while, depending on where the break occurred. If it occurred between the pull horse and the first pulley, it was a rather simple thing. But, if it occurred up in the hayloft, the whole rope needed re-stringing through all the pulleys and lift mechanisms. When this happened, we usually moved to another barn while the repair was being made.

The hay-making and the corn cultivation continued until the oats were ripe and ready for harvesting. This was a grand event for us youngsters. It meant that the Case (might have been a Rumley) steam-traction engine was going to be fired up and prepared for the threshing season. This also meant that we would get to hear that far-reaching steam whistle. If you have never heard one, you will never truly know the thrill. I doubt the sound can even be recorded

*A View from the Operator's Position
(The High Amana steam engine had a canopy)*

effectively, because each whistle carries a certain personality all its own. In our little village of High Amana, we actually had two steam whistles. The other one was at the sawmill. Each had its own distinguishing tone.

The steam engine was backed up to the Red River Special brand threshing machine (also called a separator) and attached. The separator was pulled out of its shed and prepared for the work. The belts were checked, the parts were greased, and the drive shafts and separators were inspected. The machine was leveled by digging holes under whichever wheel was a bit higher than the others. Then the steam engine was maneuvered into its place with the long belt stretched between its giant flywheel pulley and the much smaller one on the separator. Coal or wood was fed into the engine's firebox to maintain a head of steam within the required pressure range. The engineer pushed a lever and the pitman slowly started churning until the large pulley reached its required RPMs. The separator boss (he was the man in charge of the threshing machine) checked the RPMs of a certain pulley on the machine, and thus the steam engineer would know the required throttle setting, steam pressure, etc. The harvest now began in earnest.

*Threshing Machine, also called Separator
(similar to the one used on High Amana farm)*

During this same time men in the field were cutting and bundling the oats or wheat. This was done with a machine called the binder (see picture on next page). This machine looked like a giant windmill as it came across the field. A seven- or eight-foot-wide swath of grain was cut with a sickle bar (much like that on the mower described earlier) mounted on the front of the grain table. As the machine moved forward, the windmill-looking apparatus revolved so that the standing grain was gently bent backward over the sickle bar and cut. The grain head would fall neatly upon a moving canvas belt. The belt transported the cut shoots of grain up into the bundling mechanism where it was tied with a single strand of twine by an ingenious invention called the knotter. Each bundle was caught in a tripping cage. When enough bundles accumulated to make a shock, the operator tripped the cage, leaving seven or eight tied sheaves of grain lying neatly on the stubble of the cut field.

Binder

Other men and boys followed the binder and set up shocks of grain. These shockers, as they were called, picked up a bundle in each hand. With the head of the sheaf pointing upward they stood one against the other with the cut end of the sheaf jammed into the stubble of the field. Five or six more bundles were thus placed around the original two. Finally, the shocker picked up the last bundle in the trip pile, placed it over his other arm, and gently bent the grain end downward, forming a sort of a fan. This last sheaf of grain was laid across the tops of the standing bundles, thus holding the shock together and serving as protection from rain. And so, one after another, shocks appeared throughout the work day. There was nothing prettier than a neatly-shocked hillside of grain against a blue and billowy late afternoon summer sky. Now such a picture can only be found in those fields still being tended by Old Order Amish farmers. A trip to the Amish settlement near Kalona, Iowa (about 30 miles south of Amana), during the harvest season is well worth the effort.

Later, when most of the grain had been shocked, the hayracks were retrieved from various locations throughout the farm complex and employed to deliver the bundles of grain to the threshing location. Each teamster would guide his team and wagon along a row of shocks. A "pitcher" with a long-handled pitchfork threw the bundles onto the wagon, and the teamster would stack them in an arrangement which made for a large load and easy unloading at the threshing machine.

Each wagon had a gallon water jug, usually wrapped in several layers of water-soaked gunny sacking to keep the water cool, for the teamster and the pitcher to quench their thirst. Some of the older field hands would ask the teamster on his next trip to bring a cold beer or soda pop from *Onkel Willie's* store. On some really hot days the farm manager would bring out a 10-gallon milk can filled with iced lemonade for the workers. That was a special event! Another special event was when rain was in the offing for the next day. The manager would request we work into the evening, and sandwiches would be included with the lemonade.

The teamsters would often participate in unannounced contests of who could load the largest and best looking load of bundles. The "perfect load" was stacked so the corners looked virtually square. The load was topped with a shallow dome of bundles over the whole affair. A properly loaded wagon was easy to unload at the threshing machine, and over the years a teamster usually got the hang of it.

Every once in a while one of the teamsters made some extra "refreshment" stops along the way. The refreshment in this case was *Piestengel*—a rhubarb wine which was the downfall of many a young Amana man in those last days of the commune. I will tell you about wine-making later, but suffice it here to say that the fermentation process had a coincidental maturation point with the harvest. About threshing time, the wine was still milky in color but had reached the "barely drinkable" stage. It also packed quite a wallop and was wicked stuff. In addition to being the ruination of many a good, young Colony man, it was the nemesis of many a girlfriend, wife, mother, father, farm manager, and just about anyone associated with the person hooked on the vile concoction. This young wine was given the nickname of "Green River," which alluded to one of the unpleasant side effects of drinking it.

The first law of *Piestengel* states: the equilibrium of a person is inversely proportional to the amount of *Piestengel* consumed. As the day went on, one could notice a discernible change in the attitude of certain loads of bundles and the teamster himself. Though he may have been capable of loading a perfect load in the morning, by late afternoon after numerous "refreshment stops," the loads became tangled, lopsided messes. In some extreme cases, the hayrack itself might tip from the running gear of the wagon.

One teamster who was under the circumstances heretofore described, and whose name will go unmentioned, dumped his loaded wagon precisely at the point in the road directly in front of the farm manager's house. To this day I am convinced it was done with complete forethought, albeit under the influence. The load was dumped at exactly the time of the day which ensured that the team and wagon would not be available again until the morning.

Let me point out here and now that the incident above was not the norm. Most of the workers were diligent and hardworking men. Even though most would have a drink of wine or beer, it was not usually done to excess. Threshing was hard work for the most part and required the workers to perform their chores in a safe manner. There were many belts and pulleys on the separator and steam engine which could cause great harm to one who scoffed at the dangers.

The tray into which the bundles were pitched as they first entered the threshing machine was itself quite menacing. It consisted of a chain or canvas belt which moved the bundles into the hacking, tearing teeth at the throat of the separator. These teeth cut the string on the bundles and generally did the first part of the job of separating the grain from the stalk of oats or wheat (much like flails used in earlier times). Then, deep within the bowels of the dust-coughing monster, screens shook, shafts turned, shakers vibrated, blowers blew, and augers elevated in such a manner that finally the straw and chaff blew neatly from a 12-inch-diameter discharge pipe onto the straw stack, and the grain in another direction via an auger.

The stacking pipe was a 25-foot-long conduit connected to the top rear of the separator. At the end of this tube was a 45-degree adjustable elbow. The whole tube could be set to oscillate back and forth, depositing the straw in a crescent-shaped stack neatly piled by two men called "stackers." The kernels of oats or wheat, having dropped through the screens to the bottom of the separator, were gathered in a trough area of the machine and elevated up a smaller diameter pipe into one of two wagon boxes.

My first job as a part of the threshing crew entailed leveling the grain in the wagon box. Once a wagon was loaded, the spout was switched to the other empty wagon. An older boy with a team of horses (some years later, a Model H International tractor) would hook up to the loaded wagon and take the load to the granary building (see map, page 21). There, old Mr. John Nagel guided the front wheels of the wagon onto an A-framed cable hoist. A small gasoline engine powered a belt-driven speed jack, which powered the hoist and grain elevator.

The speed jack was nothing more than a series of gears (speed reducers) and clutches to allow the operator to direct the power at either the elevator or the hoist. The power was transferred from the speed jack to the elevator and hoist via one-inch-square drive shafts equipped with universal joint adapters to allow for misalignment and slight movement between the units. Here again, one had to be aware of the danger. The shafts were not protected with safety shields. Clothing could easily become tangled in the shafts or gears if one was not extra careful.

The hopper of the elevator was lowered behind the wagon so that when the wagon's tailgate was opened the grain would drop into the hopper. The elevator carried the grain up its slope to a hole in the end of the granary building at the very top. Inside the building a horizontal conveyor carried the grain to its appointed holding bin. Holes spaced in the bottom of the conveyor allowed the grain to be dumped at various places.

The High Amana granary was a three-story building. The horizontal conveyor was mounted just under the peak of the roof. All floors were similarly arranged with a walkway running through the center of the building with five or six bins, five feet in height, on each side. A system of wooden chutes with sliding stopper boards facilitated moving the grain to bins on the first and second floors.

I am told that much earlier (before portable elevators) the grain was stored in large Bemis brand cotton sacks filled at a stationary flailing mill located just south of the sawmill. These bags were then transported and stored in the granaries.

Work on the farm was certainly not limited to sowing and harvesting. The farm at High also maintained large herds of pigs and cattle. A dairy herd supplied the kitchens with milk and other dairy products. The care of these herds required their own foremen and laborers and were operations all to themselves. At peak times, labor was traded between the different departments.

Late summer, when the river was low, the teamsters were ordered to bring sand from the bars along the river. Wagons, with low-sided boxes, were loaded with sand at the river's edge. The sand was brought back to the farmyard and stockpiled for use in making concrete for various projects on the farm and in the community. This activity was a lot of fun for us, as we would go swimming in the river while the wagons were being loaded. Every once in a while a teamster would uncover a nest of snapping turtle eggs, and that would get our undivided attention. In those days, the Iowa River was a relatively clear stream and provided clean sand for construction purposes. This was before "fence-to-fence" farming and the advent of farm chemicals. It was also before the construction of the Coralville Reservoir 25 miles downstream. The reservoir created a vast silting basin in the Iowa River Valley through the Amana lands. As a result, the once pristine sandbars and clear waters are now darkened with silt from the eroded farmlands upstream.

As summer came to a close, that meant I would be going back to school. The farm work continued, though. The first frost meant that the corn wagons needed to be made ready. The regular wagons were fitted with side boards, making them four feet deep. Bank boards were also added to the left side of the wagon so that the teamster husking the corn could throw the ear of corn against them and bank them into the wagon box.

Days grew colder as the corn harvest progressed. Out in the forests surrounding the town, hired wood choppers had been at work for most of the spring and summer. Some of the woodsmen lived in small shanties in the forest. Sometimes we youngsters would visit them and listen to the stories they told.

The fallen trees were cut into logs. From the larger limbs, eight-foot fence posts were split and stacked into teepee-like stacks. Other parts of the trees provided cordwood. A cord of wood measured eight feet long, four feet wide, and four feet high. Cords were stacked throughout the forest waiting for the first frost or snowfall when the teamsters would head into the woods, load them on wagons or sleds, and haul them back to town.

Each family was given a quota of wood to carry them through the winter. At this time of year, a "saw gang" would travel from house to house sawing the four-foot lengths into lengths that would fit in the stoves. On a cold winter's morning you would wake up to the unmistakable ringing of the circular saw cutting through the frosty cordwood—zing, zing, zing—and be reminded that all this cut wood needed to be split, moved into the woodshed, and neatly stacked. There it rested for the year, drying out to fuel the fires for the following winter.

The farm never rested. It just kept rolling along from one season to the next.

Chapter V—The Sawmill

Across the street from our house, and a little east, stood the village sawmill. All winter long the yard to the north of the mill had been filled with logs brought in by sledges from the timbers around the village of High. The highland timber provided hardwoods such as white oak, hackberry, elm, hickory, and cherry. From the lowland, or river bottom, came the softwoods such as cottonwood and basswood, but it was also the source for walnut, a hardwood.

The mill provided the lumber for the various shops in the town. Walnut, cherry, and basswoods were used for furniture building, cottonwood for the repair and upkeep of barns, hickory for hammer and tool handles of all sorts, cross-grained elm for wagon tongues, oak for the wheelwright shop, and so on.

High Amana Sawmill, circa 1880
(looking northeast from the carpenter shop)

The above picture, although not very clear, shows some of the other enterprises in the village at the time. The sawmill is the large building to the left of the chimney. The building to the right of the smokestack is the tannery, with living quarters above. The tannery operated only for a short period as far as is known. The tannery's water supply came from cisterns in the hillside of the vineyard to the right edge of the picture. Evidence of one of the cisterns can still be found today. The original water supply pipe was discovered in the 1960s while the writer was building a small garden shed on the property to the east of the tannery. Notice the hayrack and team of horses in the foreground. The building immediately in front of the chimney housed a stationary threshing machine powered by the sawmill's steam engine. Also of note are the vineyards covering the south slopes to the north and east of the sawmill.

The sawmill was built into the hillside. Its north wall laid about 200 feet south of what is now designated as G Street, the main street running in front of the store and leading up to the cemetery east of town. The area between the street and the mill was used for log storage. The second floor of the mill was the area of operations and laid perhaps eight feet lower in elevation than G Street, providing a downward slope to the sawmill and the storage area in front of it.

Horses were used to drag the logs to the door of the mill. The team was spotted at the butt end of a log. The teamster then laid tongs (it looked like giant ice tongs) at the end of the log, with one tooth on one side of the log and the other on the opposite side. The device was attached to a double tree, which in turn was connected to the team's harnesses. The teamster gave the "giddy up" to the team and the giant tongs closed with its two hooks sinking into the log so that it could be dragged to the door of the mill.

Two men with large cant hooks rolled the log onto timbers fastened to the floor of the operating room. There were five or six of these 12 x 12-inch timbers placed perpendicularly to the large overhead north doors at about five feet apart on centers. The other end of them reached southward to a point level with the bed of the mill's moving carriage. The log was rolled on these timbers to the carriage.

Now came time for the most important job of the sawyer. He had to make the decision of how the log should be cut. Consideration was give to the shape, size, and straightness of the log to determine how to get the most lumber out of the log, and also the best grain pattern needed for the many ways the wood would be used later. The sawyer told the cant hook men how to position the log. Once this determination was made, the log was clamped into place on the carriage and the first cut was made.

The carriage with its log was placed in motion and began to travel westward on two rails that extended the length of the mill. The head of the log moved slowly toward the giant circular saw, which revolved on an arbor driven by the steam engine located on the lower floor. Another set of rails lay parallel on the north to the carriage rails and continued through a west door to a trestle outside the building. On that set of rails was a simple four-wheeled cart used to carry waste materials (slabs and the like) away from the operating floor. The waste was dumped over the side of the trestle and later sawed for firewood. The finished boards were temporarily piled in a storage area on the mill floor.

The waste pile also contained smaller dimensioned lumber which was not really suited for building. Slabs were thrown off the south side of the trestle and the smaller pieces off the north side. This later was cut up for kindling wood. The north side was a storehouse of material for the young boys of the village. Here we would find material for building all sorts of things. Long two x twos made excellent stilts. Shorter, flat pieces could be utilized in the construction of bridges over creeks, a tree house, or for building a bathing and diving platform on the bank of the Mill Race south of town. The projects were never ending and provided a vocational education to be envied by most modern schools—an ideal playground for a future carpenter, millwright, construction worker, or—in the author's case—civil engineer.

The fireman at the mill had the longest day. He started the fire for the steam boiler very early in the morning so that work at the mill could begin by 7 or 8 a.m. The boiler room on the south side of the mill was a story-and-a-half tall. To the south of the boiler room stood the giant metal smokestack. As young boys, we were not interested in the more technical aspects of the mill, such as how much horsepower the steam engine had. However, research indicates it might have been 15 horsepower, but with significant torque. The engine drove a line shaft mounted to the ceiling of the boiler room. The steam engine powered a large flat belt running upwards to another large pulley on the main line shaft. This main shaft was mounted on bearings and ran directly to the arbor for the circular saw.

Other pulleys on the shaft provided power for operations such as a small sizing mill. Another pulley drove the clutch mechanism for the movement of the carriage table. This was a complicated arrangement of belts, sprockets, and pulleys whereby the sawyer, by means of a long operating lever, could control the back-and-forth movement of the carriage as well as its speed. The operating lever was mounted on the wall near the saw head at about the center of the south side of the upper floor of the mill.

The mill was a noisy affair when in service. Fastened to the floor at the foot of the sawyer was a loud, foot-operated claxon horn. A simple series of blasts from the horn got the attention of the people performing the various operations of the mill. One blast for the boiler room, two for the waste handlers, three for the log feeders, and so on. When the sawyer detected a drop in the power of the mill, he honked the horn to get the fireman's attention. The attendant below would look up to an open window by the sawyer and await instructions, usually given by hand signals. I recall some mills in other Amana villages had a simple horn-and-pipe arrangement, like on steamships of old, by which the sawyer and the fireman could communicate with each other.

The sawmill provided untold hours of excitement and entertainment for the boys in town. In the early spring we awoke to the sharp blast of the sawmill steam whistle and sprang out of bed knowing that this was to be the start of a great day or week. It would be a time filled with enough new sounds and sights to keep us out of trouble for hours on end and give us more knowledge than even school could provide. It showed us how things operated, how power was transmitted, and how people worked as a team.

The mill, idle for the most part of the year, also provided an ideal place to play hide-and-seek. Groups of young boys and girls would spend hours hiding in its secret places. A balcony-type attic above the operating floor held an especially scary object. Not all the players would have the courage to hide there, especially the girls. For some reason or another, the fathers of the village decided that the attic should be the place to store a gray coffin. I was told later that this coffin had been used by the state of Iowa to ship a dead body from the mental institution at Mt. Pleasant. I do not know whether this was true or not, but it surely added to the story. One time a couple of us went up to the attic, mostly on a dare from each other, to open the coffin. We approached it very apprehensively and with the utmost trepidation. I cannot remember who finally had the courage to open the lid, but no sooner was it opened than slammed shut. With wide eyes, the brave but totally scared boy said, "There's a hand in there." We do not know what he saw, but it was years until we visited that attic again. Of course, the coffin was empty. It was used later in a funeral for an old hired hand with the nickname *Zwiebel Franz* (Onion Frank). I recall on that occasion the carpenter shop was used as a makeshift funeral home where Franz's body could be viewed.

Chapter VI—The Carpenter and Wagon Shops

The carpenter shop was located on the south side of the alley behind and due south of the general store. It was a simple story-and-a-half structure measuring about 30 x 50 feet with a walk-in door on the west side and a double door on the east. Attached to the main building was a smaller building containing a five-horsepower single-cylinder gas engine, a well pit, and grinding wheels for sharpening tools and such.

This shop was awe-inspiring for a young would-be engineer. The gasoline engine provided the power for the line shafts, pulleys, and flat belts that ran throughout the shop and drove the various woodworking machines in the building. The carpenters, Messrs. Pitz and Murbach, could set any of the machines in operation by shifting levers and belts.

Lightning Balanced Engine

Upon entering the west door one first encountered—on the left—a belt-driven wood-drilling machine. An unusual wooden pulley drove this machine. The pulley itself was about 10 inches in diameter, 24 inches tall, and mounted upon a vertical axle shaft. The pulley was attached to a lever system by which the operator could move it up and down, allowing the arbor holding the bit to drill through the wood.

Just past the drilling machine on the north wall set a three x 16-foot table that moved on rollers. The rollers were mounted on a support system fastened to the wood floor and allowed the table to move in a horizontal manner toward the east end of the building. A 20-inch-diameter circular saw was located at about the center of the support system so that a long board, resting on the table, could be ripped to the desired width as the table was pushed toward the saw blade. There were no safety devices guarding the whirling blade. Boards could be clamped to the table with wooden screw-down devices, allowing the operator to keep his hands away from the blade. With this saw the carpenter fashioned wagon tongues, reaches, and various other items too long to be handled on the smaller table saw to the right of the large rolling table.

To the right of the entrance on the west side stood a large band saw. It, too, was powered with belts and pulleys, as were all the machines in the shop. With this band saw the carpenter could "swing" circles of up to 24 inches. Parts for the rims of wagon wheels were sawed here, along with other curved items, such as the rough shapes for wheelbarrow and axe handles.

East of the band saw, near the center of the main shop, stood the 20-inch joiner. It was a massive machine which I had the chore of helping move later in life. The large joiner was fitted with a suction fan that carried the shavings through a pipe to the little covered shaving shed at the northwest corner of the building. A little farther east stood the giant cast-iron, potbellied stove. It was fired mostly with wood scraps and provided a cozy place for a young lad to stand and watch the carpenters at work in winter.

Two workbenches ran along the south side below the windows. The southern exposure provided good light for some of the more exacting projects the carpenters faced. Each carpenter had his own tool cabinet mounted on the wall. These cabinets contained their most prized possessions: hammers, chisels, braces, wood bits, various types of hand saws—cross, rip, cabinet, back, coving—and wooden hand planes of various lengths and sizes. A four-door cabinet stood between the two workbenches. Behind the upper two doors were the molding planes which, themselves, were works of art. There must have been at least a hundred or more, beautifully varnished and worn smooth from the workman's hands. Each had its distinctive blade pattern to shape the moldings of picture frames, edges of tabletops, and the like. They varied in width from three-quarters of an inch to two inches. A complete assortment of nails, screws, hinges, sandpaper, and other building materials were stored in the lower half of the cabinet.

A large wood lathe was situated in the southeast corner of the building. The carpenters used this machine to turn neck yokes for the horses, wooden tools for the community kitchens, shovel handles, and other necessary wood items of the community.

The carpenters had another duty. They were in charge of looking after the water supply for the town. From the east door of the shop they could look up towards the water tank (visible in the picture on page 4) to check the level of the float indicator sticking out of the top of the tank. A low float meant that the pumps in the well pit were required. The engine was started, and the belt power system diverted, setting the pump in motion to pump the water via an underground line to the octagonal holding tank on the hill east of Geiger's kitchen house.

The carpenter shop was a wonderful place. Not even today's Internet can provide the firsthand experience of seeing something useful and beautiful crafted from a simple piece of rough wood. The diligence and expertise that went with it had to be seen, yes, even felt and smelled. The velvety smoothness of a finely finished piece of wood, the aroma of freshly-sawed oak, pine, cherry, or walnut—none of these sensations can be captured anywhere else than by standing in the shavings of the carpenter shop.

The wagon shop stood just to the west of the carpenter shop. It, too, was a simple, one-and-a-half-storied building measuring about 30 x 50 feet with a double door and a regular door on the side facing west. There was also a door on the east side to provide easy access to the carpenter shop.

I recall only a few pieces of equipment in this shop. One was a machine to make wooden spokes for the wagons. Another punched sockets for the wheel spokes. I cannot remember how these were powered. The wagon shop maintained an inventory of spokes of different lengths and diameters—thinner ones for buggies, long and thick ones for the farm wagons, and short ones for wheelbarrows. Much of the fabrication of wagon parts was done in the carpenter and blacksmith shops, but the assembly and the finishing touches, such as painting, striping, and fancy designs, were all done in this shop.

The wheelwright's finished product was a sight to behold. There the new wagon stood on the floor, sparkling with fresh paint. The box was green with little fancy yellow stripes and decorations. The tongue and running gear were a bright barn-red. Metal edges and other hardware were painted black. Where else in the world, in a tiny village, in the third decade of the

20th century, long after the advent of the automobile and gas-powered tractors, could one observe an event like this? This youngster was indeed fortunate.

Chapter VII—The Store

There is really only one way to describe the High Amana Store in the 1930s—it was an anachronism. One would expect an enterprise in a village of this sort to be rather backward. Simplicity of lifestyle and pictism were practiced by the members of the Amana community. Though the world outside already had the amenities associated with electrical energy, this was not so in the Amanas. There was no electricity until after 1936, no indoor plumbing, no automobiles (other than a few owned by the Amana business corporation), and no modern gadgets which by then had become a part of life elsewhere. The simplicity of the people of Amana in this regard was barely distinguishable from the Amish living 30 or so miles to the south. Nonetheless, here amidst this austere lifestyle was this thriving, *modern* retail business called the High Amana Store.

High Amana Store circa 1940

Before we go on with the more detailed description of the High Amana Store, let's take a brief look at the photograph above. The large double doors are the main entrance to the store. These doors, much like large shutters, were opened in the morning and latched back against the sandstone masonry on either side of the doorway until the end of the day. The open shutters exposed another door further inside providing regular entry for the customers during the day.

Notice the old-style gasoline pump at the corner of the small brick building. A hand-operated lever pumped gasoline from the underground storage tank up into the glass cylinder at the top of the pump pedestal. This cylinder was marked off in increments of 10, each holding a gallon of gas. The gas was then dispensed by gravity through a hose and nozzle into the tank of the vehicle. The brick building was known as *das Öl Haus* (the oil house), and that is exactly what it was. In it were 50-gallon drums of various oils for engines. Each drum had its own hand-operated pump.

Look closely just outside and in front of the main entrance and you will notice a concrete platform. It was there for the convenience of customers and deliverymen. The delivery truck pulled up to the north side of the structure, and its cargo was unloaded and carried down a few steps and into the store. The part of the building to the left of the main entrance was the storekeeper's living quarters. To the right of the entrance was the warehouse, which also contained office quarters for the town's farm boss.

The white sliding door you see at the end of the warehouse was a later addition. Earlier in the 1930s the area just west of this door was the location of the town's large scale. Here, teamsters brought their wagons loaded with grain to be weighed. The scale provided the first taste of capitalism for the men harvesting the grain. Now they were being paid by the bushel and realized, perhaps for the first time, that hard work and efficient work ethics meant greater compensation. The people of the Bolshevik Revolution in Russia could have saved themselves a lot of grief had they studied the history of communal Amana and how the work ethic improved once the socialistic standards of old Amana were no longer in effect!

The Great Depression of the 1920s and 1930s was a catalyst for change. It had a profound effect upon the Amana community. "The Great Change" from a communal to a capitalistic system was a product of those hard times, and neither the people nor the Amana Society itself had the funds to greatly alter those dire conditions. In spite of this difficult economic environment, the High Amana Store was an island which seemed to have a life of its own. Here in the midst of the last vestiges of communal life was an enterprise which defied much of the world around it.

In the early 1930s, the village of High Amana had perhaps two or three motor vehicles, yet was now receiving wagonloads of automobile tires on a regular basis from the rail depot at South Amana. On the shelves of the warehouse adjacent to the store one could find car radios, auto batteries, and bicycles along with their respective repair parts inventory. Shortly before

1930, for a period of perhaps 15 years, the store was a dealer for the Indian brand of motorcycles. During this period, the High Amana Store held the distinction of being the United States' largest distributor for the Schroeder Valve Company. The Schroeder company manufactured the little valves and valve caps for the inner tubes of balloon tires, along with other items for the growing automotive aftermarket.

Brunswick Tire Sales Truck
Walter Wendler, Harry A. Wendler, and salesman circa 1932

The store was managed by an enterprising gentleman named William Foerstner, whom we called either *Bruder* (brother) *Willie* or *Onkel* (uncle) *Willie*. Some readers may remember another High Amana native, George Foerstner, who was the founder of Amana Refrigeration. George was William's son, and surely gained his entrepreneurial spirit from his father.

I remember my Uncle Ray Steele telling the story of how he came to be employed by the High Amana Store. Mr. Steele was the husband of my father's sister, Emily, born here in High Amana. Uncle Ray was born in Terre Haute, Indiana, and later moved to Cedar Rapids with his father and family during the economic depression of the late 19th and early 20th centuries. Uncle Ray was quite a mechanic and studied mechanical engineering at Iowa State College (now Iowa State University). Ray always had an interest in automobiles and, back then, motorcycles, too.

One day as a young boy, around the year 1913, Uncle Ray was crossing what was then the Fifth Avenue bridge in Cedar Rapids. He was riding his motorcycle. About halfway across the bridge the chain on his motorcycle skipped off the sprockets. He got off to repair it and noticed a man stopped to watch him work. After the chain was back on the sprockets, the man watching remarked what a swift and expert job of repairing he had just witnessed. He asked the youngster if he would be interested in a job repairing motorcycles. Well, Ray said he was sort of interested, and the man then asked him to a tavern on First Street East, where he bought Uncle Ray a lemonade while he, the man, ordered a glass of beer. They introduced themselves, and the man laid the job on the table. Uncle Ray asked where this job was located. The answer was "in High Amana at the store." The man offering the job was Willie Foerstner.

*Indian Motorcycle and Operator
High Amana circa 1915*

Ray took the job. He became the store's first repairman for Indian motorcycles. He moved to High Amana, fell in love with a young clerk at the store, Emily Wendler (my aunt), and courted her for approximately 25 years. This courtship was kept alive despite job transfers, enlistment into what was then the brand new Army Air Corps (a division of the Signal Corps) during the First World War, and overseas duty in France.

After the war, Ray came home to work in Cedar Rapids and care for his ailing mother. It was not until after his mother's death that he and my aunt finally married in 1939, a marriage which lasted until Aunt Emily's death almost forty years later.

The store always held a young boy's interest. There was a distinct and pleasant aroma to the store—an aroma composed of a blend of many odors. Ripening bananas hanging from a hook in the pressed tin ceiling of the store gave out their singular smell; oranges in their wooden crate; freshly-ground coffee in the drawer underneath the big red-and-silver coffee mill; leather gloves and shoes on the racks and shelves; spices in their canisters; and large bulk boxes of Oreo cookies in their display case. The most pungent odor seeped from a pump just inside the front door of the store. It pumped kerosene from an outside underground tank and provided villagers with oil for the pre-electric lamps and cookstoves.

To the left of the main entrance, tucked into the corner of the salesroom, was the town's post office. It took up a space of no more than 100 square feet. It was divided from the rest of the store by a curved glass-and-wood partition holding the mailboxes. *Onkel Willie* was the postmaster. At a window desk in the post office section, *Onkel Willie* also filled out the store's daily sales reports, which were sent to the Amana Society main office the next morning via the *'rum Post*, which was a corporate delivery service operating daily around the seven Amana villages. It delivered interoffice memos, notices, operating funds, meat products, etc., to the various businesses of the corporation.

The sales area of the store measured approximately 150 x 40 feet. Shelves and storage drawers filled the long walls on each side of the room, with clerks working the aisles between the wall shelves and service counters. It was a general store in the truest sense of the word.

To the left and past the post office cubicle was a door which led into the private living quarters of the storekeeper and his family. The wall to the right of the door was covered from floor to ceiling with drawers and shelves. Packaged grocery items were displayed on the shelves with backup inventory stored in the drawers below. Halfway down the aisle came the hardware department, where one found folding rulers and small tools, hinges, nails, screws, bolts, etc. A break in the counter formed an aisle to the back room (receiving department) and a stairway to the attic where other inventory was stored. Beyond this aisle one found the dry goods department. It held selections of men's and women's shoes, hosiery, straw and felt hats, undergarments, overalls—just about anything one needed.

The cash register, if that's what it can be called, was positioned near the entrance by the post office. It was a stand-up wooden desk with a money drawer which could only be opened with a secret manipulation of five small finger-levers whose combination was known only to the clerks.

At the end of the counter and just in front of the register was a machine which was adored by every youngster in the community—the Yucatan brand gum dispensing machine. A penny dropped into the slot would actuate the mechanism, which finally allowed a single stick of gum to drop out. It was a marvel to every child who was lucky enough to somehow come up with a copper penny during those hard times. Every once in a while a clerk, while reloading the machine, would insert a few random sticks of licorice-flavored Black Jack gum among the sticks of peppermint-flavored Yucatan gum. Whoever got this stick of gum was entitled to one additional stick free of charge! It was a great day when one was lucky enough to get the piece of Black Jack, for you would immediately be the envy of all the other kids to whom you could hardly wait to tell the news of your good fortune.

Brother Herman Puegner in the dry goods department of the High Amana Store, circa 1930. The picture is taken looking north towards the entrance of the store.

In the center of the main sales floor (behind and left of where Mr. Puegner is sitting in the picture above) were glass showcases in which were displayed the "finer" things of the day—decorative items for the home or perhaps a set of dishes for the new housekeepers, who shortly prior to this time did all their cooking in the communal kitchens of the village. Yard goods were displayed on shelves on the west wall and stored under the counters there. A pretty display case stood in the southwest portion of the store. This case held the more valuable items, such as pocket watches, straight and safety razors, combs, hairbrushes, and other toiletry and beauty items. It was a really great day when, after working on the farm during the summer, one had enough money to buy a new Westclox dollar pocket watch from this case.

Not shown, but just to Mr. Puegner's right, is a door leading to the receiving room mentioned earlier. A large double door was located at the south end of the receiving room. Here the wares were unloaded from horse-drawn wagons delivering from the Rock Island & Pacific Railway station in South Amana or the Milwaukee Road station in Upper South. The stairway in the room led up to the attic where the seasonal items were stored. Upstairs was also the very secure room known to all the children of High Amana as *die Weinachts Stube* (the Christmas room). What a room this was! Here *Onkel Willie* would store all the Christmas-related items he received in shipments during the year.

The storekeeper and his clerks would select a date on which the store would be converted to its once-a-year Christmas splendor. On the evening before the date, a "tree" was placed on the fancy showcases in the center of the sales floor and decorated with garlands of popcorn, tinsel, and wax candles in gold and silver candleholders clipped to the branches of the "tree." The "tree" itself was fashioned from a broomstick and fastened to a base. The broomstick was drilled with holes at various levels. Pine boughs cut from pine groves of windbreaks in the villages were stuck into the holes, starting with longer boughs at the bottom and gradually getting shorter toward the top of the tree. One last vertical short bough was stuck at the very top. Small pine trees were not available from the hardwood forests surrounding the villages; therefore, similar *ersatz* Christmas trees could be found in many Amana homes.

A blanket of white cotton batting surrounded by a tiny handcrafted green fence was placed at the base of the tree. Inside the fence were miniature figures of animals—sheep, horses, cattle, hens, and rabbits—making a pastoral winter scene. The sales counters held displays of Christmas goodies—toys, candies, special fruits, seasonal items, and things one did not often see in this tiny, Depression-era village.

We children would become very restless as Christmas approached. *Onkel Willie* and his clerks took much delight in keeping us in suspense as to the date when the store would be transposed into the holiday wonderland. Actual preparation for the long-awaited event took place during the evening before the announced date. We awaited the dawn of the next day when all the youngsters of the village rushed to the store to stare with open-mouthed wonderment at the beauty of it all, especially the display of a few new toys. All our conversations with friends now centered on what we had seen. Our hope against hope was that on Christmas Eve we would have the good fortune to find at least one of the items below the tree in our home. The High Amana Store at Christmastime was an enchanting place. There were also special items displayed for the older folks of the community—a fancy meerschaum pipe for a grandfather, a flannel nightgown for a wife, a tiny bottle of perfume (always Evening in Paris) for someone special. More mundane, but just as eagerly anticipated by every family, was a barrel of herring in salt brine from which a special salad called *Herring Salat* (herring salad) was made by the homemakers of the community. This salad was made of cubed fillets of herring, apples, onions, and fresh cream and was always a tasty holiday tradition.

Who would think that hidden in a tiny, German-speaking village in Iowa one could find an enterprise as interesting and full of surprises as the High Amana Store.

Chapter VIII—The Church

The history of the Amana Church Society is well documented, going back to the late 17th century pietism movement in Europe. Much has been written regarding the church. The reader who is interested in a more detailed history is urged to search out the bibliography available through Amana Heritage Society and Amana Church Society archives. For simplicity's sake, let it suffice to say that in 1714 an ordained minister of the Lutheran faith named Eberhard Ludwig Gruber founded the Amana Church, technically named The Church of the True Inspiration. Gruber became disenchanted with the ritual of the Lutheran church and adopted the philosophy that a believer needs no intermediary (priest or minister) between himself and God. He further espoused that the word of God could be heard spoken through certain persons who maintain a very close relationship with the Lord, just as it was in biblical times. People possessing such a relationship were described as *Werkzeuge* (instruments) of the Lord. According to the belief of this new church, these *Werkzeuge* were able to communicate the Word of God either orally or in writing.

Throughout its history, The Community of True Inspiration had several *Werkzeuge*, or instruments. E.L. Gruber was not a *Werkzeug*, but his son, Johann Adam Gruber, was. Johann Adam immigrated to America a century before the people of Amana did; although leaving the faith, he remained in close contact with the community in Germany.

Christian Metz, the *Werkzeug* responsible for the move to America in the mid-19th century, and a lady named Barbara Heinemann Landmann carried on the work of inspiration here in the Amana community until Landmann's death, around the turn of the century. Since then the church has had no recognized *Werkzeug*.

The death of Landmann left the church in a state of limbo. The voices of the *Werkzeuge* no longer echoed in the chambers of the church meetinghouses. Instead, the previously spoken testimonies, filling volume upon volume, were read aloud by the presiding elder. This practice still remains a part of the services of the present Amana Church.

In the year 1931 the members voted to separate the church from the economic affairs of the organization. And so it was in 1932 that "The Great Change" was initiated, and the commune ended.

Everyone in our town attended church services, and the vast majority attended services three or four times a week. When I was a child we had three services on Sunday. The first was the 9 a.m. service, followed by a 1 p.m. service, and finally a 7 p.m. evening prayer service. At one time there was also a midweek service which may have continued into my very early childhood, although I cannot recall.

I attended Sunday school, which was held in what was the former *Kleine Saal* (little meetinghouse) across the street from the main church. I recall going to the school at a very young age. This was probably due to the fact that my Aunt Emily was one of the teachers. Our Sunday school was closed in the early 1930s, perhaps because of the lack of teachers. At any rate, I began to attend the regular church services instead and, a few years later when auto travel became more available, I attended Sunday school in Middle Amana until confirmation at the age of 15.

Four *Vorsteher* (elders) rotated conducting the church services in High Amana. The very young boys, such as me, had to sit on the very first bench in the church, right in front of the elders. I was a lad of six at the time, and I thought the elders were ancient and very, very stern. The elder who sat right in front of me and faced me directly was my grandfather, Arno Wendler. My feet dangled from the hard pine bench, unable to reach the floor. Words were read from the Bible, big words, hard to understand, and even more so because they were in German. The hymns, also in German, were long and sung very slowly. Each elder would give a homily at the end of the scripture reading. Some, like my grandfather, kept it "short and sweet," but others had a tendency to go on and on, and in German, of course.

Now, it is hard for a little boy to sit still on a hard bench too high for his feet to touch the floor and with a hard backrest that for him was actually a headrest because it was so high. It was a losing effort to keep from shifting around and swinging my legs and generally making it known that I had had enough religion for one day. The worst part came when I realized what I was doing and had to look at my grandfather's wrinkled brow and squinting eyes. He spoke not a word, but the message was loud and clear: "QUIT DOING THAT!"

The High Amana Church was the smallest of the churches in the seven Amana villages.

I started out early carrying a *Psalter-Spiel* (German hymnal) to church. The hymnals and Bibles were handed down through the generations. Most of them contained little celluloid bookmarks which were given at special occasions to the children of the Sunday school. Each marker had a simple verse and scripture passage printed on it. Many of these were as old as the handed-down hymnal itself. I don't believe anyone ever taught us this, but sooner or later each youngster found that if the celluloid marker were placed on the palm of the hand, the heat of the hand would cause it to curl. Then you could place it on your other palm and begin the process of curling it in the other direction. This kept our attention for minutes at a time until that awakening moment when you realized what you were doing and then, BAM! I would be looking right into the eyes of my grandfather again.

The reader should not think that every youngster turned into a heathen. For example, we wondered why the elder Brother Puegner often had tears in his eyes when he read a verse from a hymn. Eventually I came to realize that things of beauty can have an influence on us, and now I find myself equally moved when beautiful words or music touch my senses. No, we did not turn into little heathens. Our church (and the Sunday schools) did a wonderful job of teaching us, though for the most part we were unaware that such a learning experience was actually taking place. We were instructed, not only in the basic story of the Bible, but at the same time were also taught a second language for which I have been grateful all my life. Early on we were able to read the old-style German print in our hymnal and eventually followed along singing the songs and reading the verses. We were taught the simple way of worship. We were taught to respect our elders. We were taught to attend church. We were taught the love of Jesus Christ, and this still remains the central purpose of our church.

Earlier, I mentioned the first funeral I attended. Actually, I did not attend the church service, but I did attend *Geiger Opa's* interment at the High Amana cemetery. You will recall that he was the elderly man who lived with his wife on the main floor of my childhood home.

Geiger Opa died early on November 3, 1935. I had just turned four years old, and this marked the first indelible event that I can recall from my childhood.

I remember the solemnity within the household during the time before *Opa's* death. I did not understand it at all. Death was a thing unknown to me, and not until much later would I begin to understand its significance. There was a lot of "shushing" and reminders not to play too noisily or run up or down the stairs. The attitudes of my parents were quite mysterious. Then one morning Dad returned upstairs after looking in on *Geiger Opa* and *Oma*. He just nodded to Mom as he came into the small kitchen area. After a few hushed words with her they sat down and tried to break the news to me. How do you explain death to a little boy who has no understanding of the concept? They did their gentle and "churchly" best, I am sure.

Of course, as a four-year-old I do not remember the particulars of that day. It is only by subsequent observance of other funerals that

I can reconstruct the events of my first experience with death.

Dad went across the street to the store and broke the news to *Br.* (abbreviation for *Bruder,* or brother) *Willie*, the storekeeper and an elder in the church. The store had the only telephone in town. Other people were called and given the message—Dr. Hermann in Middle and the undertaker, Mr. Eichacker, in Homestead. The two came to High Amana later that day: Dr. Hermann to sign the death certificate and Mr. Eichacker to embalm the body. By now the news had reached everyone in our little village.

Br. Willie, also a church elder, was appointed to conduct the service. He came early that day to visit with *Geiger Oma* and members of her immediate family. A prayer of condolence and courage was spoken, and then the arrangements for the funeral service were discussed. Hymns were selected from the *Psalter-Spiel.* Favorite passages, perhaps marked by *Geiger Opa* in his personal Bible, were selected for the service. After this, the elder offered a blessing and a benediction and then left to prepare for the funeral.

Br. Willie then returned to his store where he would have prepared the announcement for the funeral, written on a black-edged, half-sheet of paper. On this pre-printed form he would fill in the appropriate blank spots with the name of the deceased and time of the service. It would be written in German script and read: *"Die Leichenversammlung für den lieben vestorbenen Mit-Bruder und Eldesten, Jacob Geiger, wird am 6ten November am halb uhr nachmittags in Amana vor der Höhe gehalten."* ("The funeral service for our dear departed brother and elder, Jacob Geiger, will be held in the afternoon of November 6th at 12:30 p.m. in High Amana.") *Br. Willie* phoned this information to the other stores in the seven villages where similar notices were posted.

The men at the carpenter shop got the casket ready for delivery to the home and the rough box for the grave. Because of the frozen November ground, a wagonload of corncobs was hauled to the grave site. The pile of cobs was set on fire, thawing the ground so that the gravediggers could open the grave the next day. After the grave was dug, the rough pine box was lowered into it. A heavy wood plank lay flat on either side of the opening. Two crosspieces spanned the grave to support the casket above the open grave.

The village now became a beehive of activity. *Geiger Opa* had been the senior elder, a distinguished position. Therefore, many mourners were expected from the other villages. The women of the town adopted a frenzied cleaning mode, because once the services were over, relatives and friends from the colonies would be invited for lunch and fellowship in the homes of the residents. The baker baked an extra oven of bread on the morning of the funeral to serve all those attending. The butcher made sure to have enough sausage and ham for the occasion. The ovens in the kitchen houses (most of which were still intact in 1936) were fired up to bake the cakes (feather, angel food, and chocolate) and pies (plum, peach, and cherry.)

Later that day my father came to me and asked me to go with him. We went downstairs into the Geigers' apartment. There, at a south window of the house, was this long, walnut-colored thing that had never been there before. Dad carried me over to the mysterious brown box with half its top open. In it was *Geiger Opa*, sleeping as far as I could tell. He looked quite at ease, and everything seemed okay to me.

During that day and the next, more of *Geiger Opa's* immediate family arrived. Most were from the village of Homestead. His two grandsons, Otto and George, came to pay their respects. Both of these boys were a little older than me, and we played while the parents and family visited. The next day, many people came to pay their last respects. Each gave condolences to the family and visited for a short while. All conversation was in German.

Someone, probably Aunt Emily, stayed with me during the time of the actual church service, because I was too young to attend. Later experiences again allow me to reconstruct this service.

Villagers on the way to High Amana church service, circa early 1900s

Each village was represented by at least one elder, and most sent more than one. The High Amana Church was the smallest of the seven Amana churches. Many who would have ordinarily attended stayed away because of the limited available seating. Even so, the congregation filled the church to overflowing. Extra benches, brought from the Sunday school, were set up in the hallway outside the main sanctuary. Still others were seated on chairs in the small apartment at the west end of the church building.

Br. Foerstner began with the invocation: *"Lasset uns dann, in Gottes namen, die Leicheversammlung für den lieben verstorbenen mit-eldesten, Br. Jacob Geiger, anfangen mit dem Lied, 'Trinkt nur getrost den bittern kelch zum sterben.' Es stehet auf der Seite 953."* ("Let us then, in God's name, begin our funeral service for our dear departed brother and elder, Jacob Geiger, with the hymn....")

The congregation sat on the benches facing the elders, men seated on one side of the church and women on the other. Following the hymn, an appropriate testimony of a *Werkzeug* was read, after which the congregation knelt for prayer. Still kneeling, everyone would sing three verses of a second hymn. The presiding elder read a single line at a time from the stanzas which the kneeling congregation would repeat in song. When finished, all rose and sat. Then the presiding elder announced the scripture passages to be read. All present turned to the appropriate chapter, and one by one the verses were read. The presiding elder would read the first verse of the chapter, followed in order by the elders and then individuals of the congregation until the chapter was completed.

Now followed the *Erinerrung* (remembrance or homily) by the elders, beginning with the presiding elder. Each would touch upon the life of the deceased, mentioning his good points (in some cases the bad) and how all of us would one day face the Creator. All comments were tied to the scripture passage. One by one the other elders followed with their own commentary that could last for up to five minutes, depending upon the oratory skill of the elder and respect for the deceased. Reading of a Psalm followed along with some more words by the presiding elder. The closing hymn was announced and sung with everyone standing for the last two verses. The service closed with a benediction. Generally, a funeral service lasted an hour to an hour-and-a-half, depending upon the respect of the deceased, the number of elders in attendance, and the length of the individual commentaries.

The women exited first, followed by the men (the High Amana Church had only one entrance to the main meeting room). All walked up the hill to the Geigers' house to pay final respects. Finally, the presiding elder would gather the family near the casket and offer a final blessing. The lid was placed upon the casket, which then was carried through the door to the black funeral wagon waiting outside.

While everyone was at the actual church service, my aunt took me to the window as the funeral wagon, pulled by a team of bay horses, arrived. The teamster was my Uncle Fred Wendler. He maneuvered the team up the driveway, turned the wagon around so the team faced downhill toward the street, and there waited for the pallbearers to load the casket.

Onkel Freddie (Uncle Fred) had taken most of the morning preparing the horses to look their very best. Each village was proud of its own funeral team of horses. The best matched and best behaved team was chosen for the cortege. Their manes were braided with delicate little braids, and their coats were shiny—as if given a light coating of wax. Hooves were cleaned and trimmed, and the white fetlocks sparkled in the snow. *Onkel Freddie* was seated on the spring seat at the front of the black, open funeral wagon that had been dusted and cleaned. It was a crisp November day, so he wore a fur coat and a hat with earmuffs down. A heavy buffalo robe covered his feet and lap. The blustery November cold would not have influence on his duties this day.

After all had paid their last respects, the casket was loaded and the procession to the cemetery began. First behind the wagon were the pallbearers. Next came the elders, two abreast, and then the male members of the family. My dad and I were included in this group even though we were not blood relatives. Following us came all the other men, and then the female members of the family. Finally, the female friends of the family fell in behind. The women all wore black dresses, with black aprons, shawls, and caps.

Two by two, the cortege crept slowly up the hill toward the cemetery. The wagon entered the west gate under the stately white pines surrounding the graveyard and proceeded down the path between the simple, identical headstones to the open grave. The people approached, surrounded the grave site, and stood quietly watching as the coffin was unloaded and carried to rest on the crosspieces over the grave. The congregation gathered around and sang another few verses of a hymn. The presiding elder said a prayer, followed by The Lord's Prayer spoken in unison by the congregation. After the Amen, all left the cemetery and went to the houses of friends and relatives for the "lunch."

Only the pallbearers remained behind. Their job was about completed, but first the casket had to be lowered into the rough box in the grave. Two ropes were threaded under the casket and held by four men, two on each side of the grave. They lifted the box off the supporting crossbeams. The other two pallbearers slipped the supports from under the casket, the casket lowered, and the body laid to rest. The gravediggers (often two nonmember day laborers) stayed behind to close the grave. The wooden planks and supports were loaded on the wagon and returned to their storage place for the next funeral.

The Amana religion views a funeral as a celebration, although to an "outsider" it might look quite somber. It taught us (and still does) to look upon death as the beginning of a great adventure—as the beginning of a new and endless life with the Creator. In keeping with the teaching of the church's founder, E.L. Gruber, the members of the church hold a strong, personal relationship with God. We know He is there and we know where we are going.

Now the service is over. All gather in the homes of the villagers—aunts and uncles, *Opas* and *Omas*, friends and relations, all sharing in family stories and in the general news of the other villages. Old times were reminisced.

The men might go to the basement for a glass or two of homemade *Piestengel* or grape wine and perhaps a cigar. The women prepared the tables with the food—potato salad, pickles, pickled yellow beans, ham, sausages, cheese, fresh bread and butter, and coffee or lemonade, depending upon the time of the year. It was a wonderful gathering where family history was passed down from one generation to the next without anyone really aware it was taking place.

As the early winter sun settled toward the horizon, the mourners took to their newfangled form of transportation, the automobile, now owned by a few of the residents of Amana. A few years earlier, horses and buggies would have been used. *Aufwiedersehens* are exchanged, and everyone is happy and at peace.

Deep inside I knew the same was true for *Geiger Opa.*

Conclusion

High Amana in the 1930s, like all of the other six villages of the Amana Colonies, was a perfect place for a young boy, such as me, to grow up. I hope that my reader has had a taste of this life and the adventures that made my early years so enchanting. Some years ago my sister, Janet Zuber, wrote a poem about our little town. Perhaps the thoughts and descriptions embodied therein provide the most fitting way to conclude my story.

High Amana

The beauty of our little town
That's here for all to find,
Is the spirit she instills within,
A precious peace of mind.

She lies beneath a healthy hill,
Atop a prosperous plain,
And joins the two in harmony,
For all who here remain.

She gives to her inhabitants,
Those gone and yet to come,
A certain continuity
Which makes their lives as one.

Janet W. Zuber, 1964